Author Bakthi Ross © 2017.08.17

For information or to order additional books, please write to

Waxwing

PO Box 373

MORAYFIELD 4506

AUSTRALIA

OR PHONE 07-54987214

ISBN 978 1 922220 36 3

The Tall House

By Bakthi Ross

A house, that was built tall and straight, went up and up.

Then it leaned to one side
but, because it was tall it
reached space.
First it went left.

Then it went right.

It curved like a wheel.

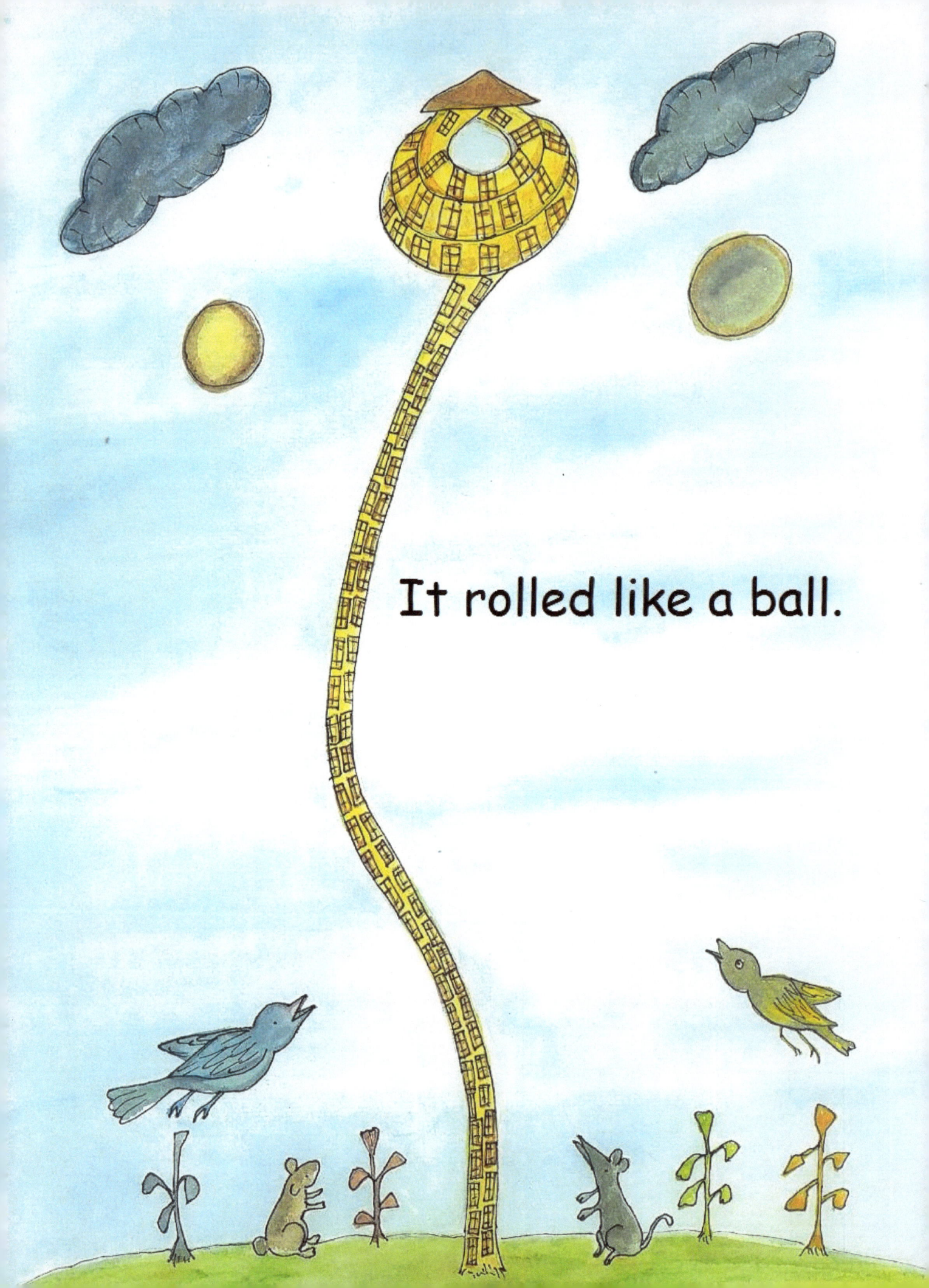

It rolled like a ball.

A globe with windows and doors.

We stuck our heads out
to look, and guess what?
We were near the planet
Mars.

Inside the house we moved
like a spaceman.

When we opened the
window our heads
stretched like aliens'
heads.

We floated outside the
house and we didn't stay
in the shape we were. We
extended like the house.

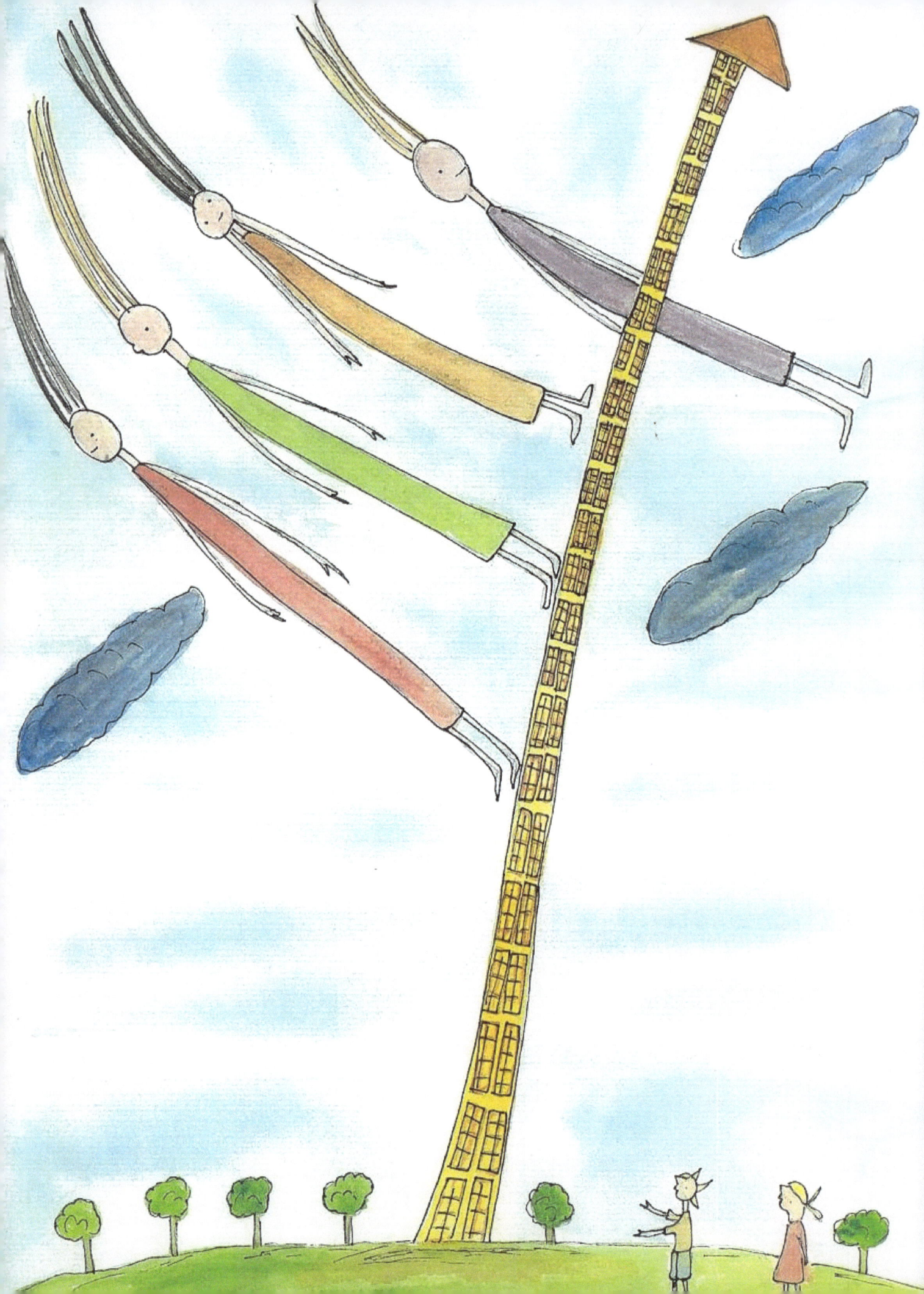

Our hands and legs grew long and our heads went straight up. We moved like a comet with our hair streaming backwards.

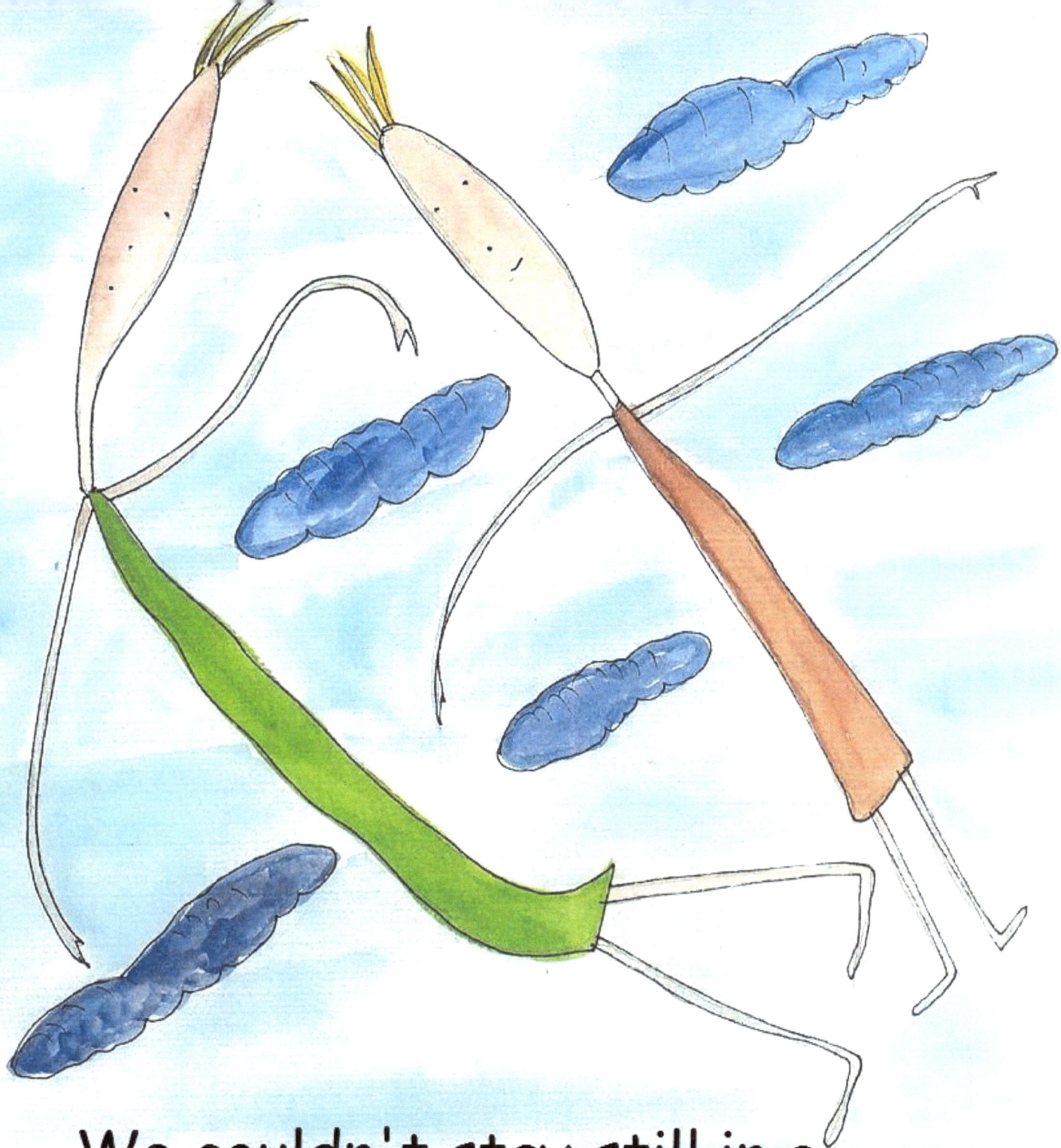

We couldn't stay still in a
house that reached the sky.
The house rotated like a
planet house.

The tall house swayed
here and there in space.

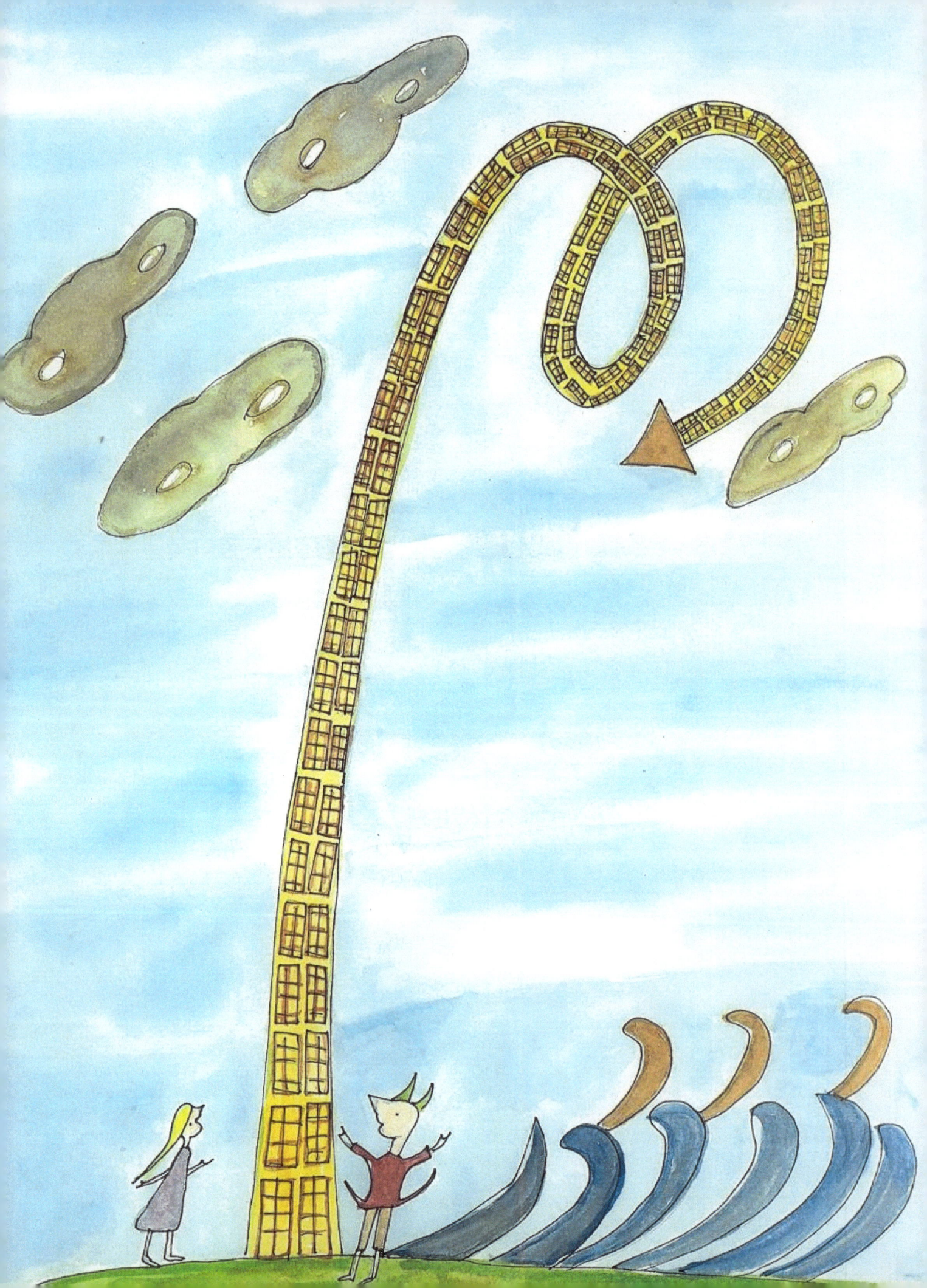

Then it rotated towards Earth.

It shrunk down to a
normal house and it
wasn't tall.

Our house on Earth is more comfortable than the house that reached the sky.

www.ingramcontent.com/pod-product-compliance
Lightning Source LLC
Chambersburg PA
CBHW052045190326
41520CB00002BA/194